This Journal No.: _____   _____

INVENTOR'S JOURNAL

Inventor Name: _____

INSTRUCTIONS FOR USING THIS JOURNAL:

<u>If you are an independent inventor, and also work for a company or other organization, first check whether any of your ideas will belong to your employer instead of to you.</u> If so, use your employer's Engineering Notebook and process.

Crease this Journal in the middle. Never tear a page of this Journal.

Write each new topic in the Table of Topics.

Start writing or drawing or mind-mapping about a new topic at a new Thoughts & Ideas Canvas. (A canvas is a left-hand-side page taken together with its opposite right-hand-side page.) Continue with canvases sequentially, signing and dating those that you filled out. Then go back to the Table of Topics, and enter into it any additional canvas numbers for this topic.

PATENT READY® INVENTOR'S JOURNAL: WITH PATENT ANALYSIS FORMS
v1.1

By Patent Introductions, Inc.
www.patent-introductions.com
www.patent-ready.com

Authored for Patent introductions, Inc. by: Gregory T. Kavounas, US Patent Attorney, MBA

Also by Patent Introductions, Inc.:

PATENT READY®: Introductory Book For Executives, Managers, Engineers & Others (2015)
PATENT READY® Engineering Notebook: With Patent Analysis Forms (2015)

LEGAL NOTICES: If working for a company, use this Journal only as directed. This Journal does not contain legal advice. Obtaining a copy of this Journal does not form an attorney-client relationship with anyone. Using this journal alone does not create any patent protection for an inventor. Rather, to obtain patent protection one should contact a patent attorney about their facts and circumstances.

TRADEMARK NOTICES: PATENT READY® and the Scroll Logo: ®️ are trademarks of Patent Introductions, Inc.

COPYRIGHT FOR PATENT READY® INVENTOR'S JOURNAL: 2015 by Patent Introductions, Inc.

ISBN: 1515308693
ISBN-13: 978-1515308690

CreateSpace
North Charleston, South Carolina

<h2 style="text-align:center;"><u>TABLE OF TOPICS</u></h2>

Topic	Canvas No.

1 ... Thoughts & Ideas Canvas ...

Topic: _____

... 1

Diagram of: _____

Signature:		Date:	
Patent Analysis Forms in page(s):			

2 ... Thoughts & Ideas Canvas ...

Topic: continued from previous: _____ / Canvas #: _____ / new: _____

... 2

Diagram of: _____

Signature:		Date:	
Patent Analysis Forms in page(s):			

3 ... Thoughts & Ideas Canvas ...

Topic: continued from previous: _____ / Canvas #: _____ / new: _____

... 3

Diagram of: _____

Signature:		Date:	
Patent Analysis Forms in page(s):			

4 ... Thoughts & Ideas Canvas ...

Topic: continued from previous: _____ / Canvas #: _____ / new: _____

... 4

Diagram of: _____

Signature:		Date:	
Patent Analysis Forms in page(s):			

5 ... Thoughts & Ideas Canvas ...

Topic: continued from previous: _____ / Canvas #: _____ / new: _____

... 5

Diagram of: _____

Signature:		Date:	
Patent Analysis Forms in page(s):			

6 ... Thoughts & Ideas Canvas ...

Topic: continued from previous: _____ / Canvas #: _____ / new: _____

... 6

Diagram of: _____

Signature:		Date:	
Patent Analysis Forms in page(s):			

7 ... Thoughts & Ideas Canvas ...

Topic: continued from previous: _____ / Canvas #: _____ / new: _____

... 7

Diagram of: _____

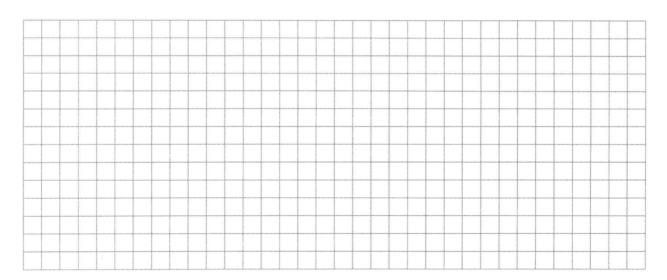

Signature:		Date:	
Patent Analysis Forms in page(s):			

8 ... Thoughts & Ideas Canvas ...

Topic: continued from previous: _____ / Canvas #: _____ / new: _____

... 8

Diagram of: _____

Signature:		Date:	
Patent Analysis Forms in page(s):			

9 ... Thoughts & Ideas Canvas ...

Topic: continued from previous: _____ / Canvas #: _____ / new: _____

... 9

Diagram of: _____

Signature:		Date:	
Patent Analysis Forms in page(s):			

10 ... Thoughts & Ideas Canvas ...

Topic: continued from previous: _____ / Canvas #: _____ / new: _____

... 10

Diagram of: _____

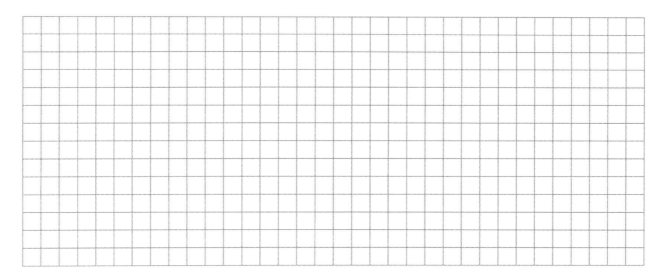

Signature:		Date:	
Patent Analysis Forms in page(s):			

11 ... Thoughts & Ideas Canvas ...

Topic: continued from previous: _____ / Canvas #: _____ / new: _____

... 11

Diagram of: _____

Signature:	Date:
Patent Analysis Forms in page(s):	

12 ... Thoughts & Ideas Canvas ...

Topic: continued from previous: _____ / Canvas #: _____ / new: _____

... 12

Diagram of: _____

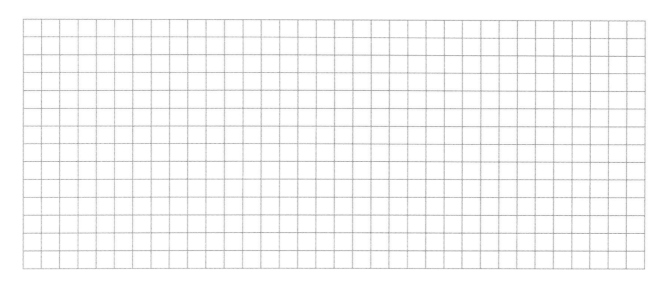

Signature:		Date:	
Patent Analysis Forms in page(s):			

13 ... Thoughts & Ideas Canvas ...

Topic: continued from previous: _____ / Canvas #: _____ / new: _____

... 13

Diagram of: _____

Signature:		Date:	
Patent Analysis Forms in page(s):			

14 ... Thoughts & Ideas Canvas ...

Topic: continued from previous: _____ / Canvas #: _____ / new: _____

... 14

Diagram of: _____

Signature:		Date:	
Patent Analysis Forms in page(s):			

15 ... Thoughts & Ideas Canvas ...

Topic: continued from previous: _____ / Canvas #: _____ / new: _____

... 15

Diagram of: _____

Signature:		Date:	
Patent Analysis Forms in page(s):			

16 ... Thoughts & Ideas Canvas ...

Topic: continued from previous: _____ / Canvas #: _____ / new: _____

... 16

Diagram of: _____

Signature:		Date:	
Patent Analysis Forms in page(s):			

17 ... Thoughts & Ideas Canvas ...

Topic: continued from previous: _____ / Canvas #: _____ / new: _____

... 17

Diagram of: _____

Signature:		Date:	
Patent Analysis Forms in page(s):			

18 ... Thoughts & Ideas Canvas ...

Topic: continued from previous: _____ / Canvas #: _____ / new: _____

... 18

Diagram of: _____

Signature:		Date:	
Patent Analysis Forms in page(s):			

19 ... Thoughts & Ideas Canvas ...

Topic: continued from previous: _____ / Canvas #: _____ / new: _____

... 19

Diagram of: _____

Signature:		Date:	
Patent Analysis Forms in page(s):			

20 ... Thoughts & Ideas Canvas ...

Topic: continued from previous: _____ / Canvas #: _____ / new: _____

... 20

Diagram of: _____

Signature:		Date:	
Patent Analysis Forms in page(s):			

21 ... Thoughts & Ideas Canvas ...

Topic: continued from previous: _____ / Canvas #: _____ / new: _____

... 21

Diagram of: _____

Signature:		Date:	
Patent Analysis Forms in page(s):			

22 ... Thoughts & Ideas Canvas ...

Topic: continued from previous: _____ / Canvas #: _____ / new: _____

... 22

Diagram of: _____

Signature:		Date:	
Patent Analysis Forms in page(s):			

23 ... Thoughts & Ideas Canvas ...

Topic: continued from previous: _____ / Canvas #: _____ / new: _____

... 23

Diagram of: _____

Signature:		Date:	
Patent Analysis Forms in page(s):			

24 ... Thoughts & Ideas Canvas ...

Topic: continued from previous: _____ / Canvas #: _____ / new: _____

... 24

Diagram of: _____

Signature:		Date:	
Patent Analysis Forms in page(s):			

25 ... Thoughts & Ideas Canvas ...

Topic: continued from previous: _____ / Canvas #: _____ / new: _____

... 25

Diagram of: _____

Signature:		Date:	
Patent Analysis Forms in page(s):			

26 ... Thoughts & Ideas Canvas ...

Topic: continued from previous: _____ / Canvas #: _____ / new: _____

... 26

Diagram of: _____

Signature: | | Date: |

Patent Analysis Forms in page(s): |

27 ... Thoughts & Ideas Canvas ...

Topic: continued from previous: _____ / Canvas #: _____ / new: _____

... 27

Diagram of: _____

Signature:		Date:	
Patent Analysis Forms in page(s):			

28 ... Thoughts & Ideas Canvas ...

Topic: continued from previous: _____ / Canvas #: _____ / new: _____

... 28

Diagram of: _____

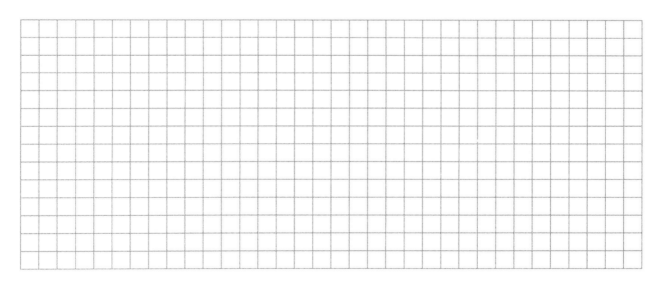

Signature:		Date:	
Patent Analysis Forms in page(s):			

57

29 ... Thoughts & Ideas Canvas ...

Topic: continued from previous: _____ / Canvas #: _____ / new: _____

... 29

Diagram of: _____

Signature:		Date:	
Patent Analysis Forms in page(s):			

30 ... Thoughts & Ideas Canvas ...

Topic: continued from previous: _____ / Canvas #: _____ / new: _____

... 30

Diagram of: _____

Signature:		Date:	
Patent Analysis Forms in page(s):			

31 ... Thoughts & Ideas Canvas ...

Topic: continued from previous: _____ / Canvas #: _____ / new: _____

... 31

Diagram of: _____

Signature:		Date:	
Patent Analysis Forms in page(s):			

32 ... Thoughts & Ideas Canvas ...

Topic: continued from previous: _____ / Canvas #: _____ / new: _____

... 32

Diagram of: _____

Signature:		Date:	
Patent Analysis Forms in page(s):			

33 ... Thoughts & Ideas Canvas ...

Topic: continued from previous: _____ / Canvas #: _____ / new: _____

... 33

Diagram of: _____

Signature:		Date:	
Patent Analysis Forms in page(s):			

34 ... Thoughts & Ideas Canvas ...

Topic: continued from previous: _____ / Canvas #: _____ / new: _____

... 34

Diagram of: _____

Signature:		Date:	
Patent Analysis Forms in page(s):			

35 ... Thoughts & Ideas Canvas ...

Topic: continued from previous: _____ / Canvas #: _____ / new: _____

... 35

Diagram of: _____

Signature:		Date:	
Patent Analysis Forms in page(s):			

36 ... Thoughts & Ideas Canvas ...

Topic: continued from previous: _____ / Canvas #: _____ / new: _____

... 36

Diagram of: _____

Signature:		Date:	
Patent Analysis Forms in page(s):			

37 ... Thoughts & Ideas Canvas ...

Topic: continued from previous: _____ / Canvas #: _____ / new: _____

... 37

Diagram of: _____

Signature:		Date:	
Patent Analysis Forms in page(s):			

38 ... Thoughts & Ideas Canvas ...

Topic: continued from previous: _____ / Canvas #: _____ / new: _____

... 38

Diagram of: _____

Signature:		Date:	
Patent Analysis Forms in page(s):			

39 ... Thoughts & Ideas Canvas ...

Topic: continued from previous: _____ / Canvas #: _____ / new: _____

... 39

Diagram of: _____

Signature:		Date:	
Patent Analysis Forms in page(s):			

40 ... Thoughts & Ideas Canvas ...

Topic: continued from previous: _____ / Canvas #: _____ / new: _____

... 40

Diagram of: _____

Signature:		Date:	
Patent Analysis Forms in page(s):			

PATENT ANALYSIS FORM

For topic of: _____ Canvas #: _____

PRELIMINARY PRIOR ART SEARCH: (See Appendix A)

Keywords: _____

Other keywords learned from early search results: _____

SEARCH RESULTS: Search date: _____

Prior art: (e.g. Patent Document #)	Seems to disclose, which is of interest to this topic:	... but not:

POSSIBLE TO PATENT: (See Appendix B)

Ideas in this canvas that do not seem to be disclosed in the prior art:

PATENT ANALYSIS FORM

For topic of: _____ Canvas #: _____

PRELIMINARY PRIOR ART SEARCH: (See Appendix A)

Keywords: _____

Other keywords learned from early search results: _____

SEARCH RESULTS: Search date: _____

Prior art: (e.g. Patent Document #)	Seems to disclose, which is of interest to this topic:	... but not:

POSSIBLE TO PATENT: (See Appendix B)

Ideas in this canvas that do not seem to be disclosed in the prior art:

PATENT ANALYSIS FORM

For topic of: _____ Canvas #: _____

PRELIMINARY PRIOR ART SEARCH: (See Appendix A)

Keywords: _____

Other keywords learned from early search results: _____

SEARCH RESULTS: Search date: _____

Prior art: (e.g. Patent Document #)	Seems to disclose, which is of interest to this topic:	… but not:

POSSIBLE TO PATENT: (See Appendix B)

Ideas in this canvas that do not seem to be disclosed in the prior art:

PATENT ANALYSIS FORM

For topic of: _____ Canvas #: _____

PRELIMINARY PRIOR ART SEARCH: (See Appendix A)

Keywords: _____

Other keywords learned from early search results: _____

SEARCH RESULTS: Search date: _____

Prior art: (e.g. Patent Document #)	Seems to disclose, which is of interest to this topic:	... but not:

POSSIBLE TO PATENT: (See Appendix B)

Ideas in this canvas that do not seem to be disclosed in the prior art:

PATENT ANALYSIS FORM

For topic of: _____ Canvas #: _____

PRELIMINARY PRIOR ART SEARCH: (See Appendix A)

Keywords: _____

Other keywords learned from early search results: _____

SEARCH RESULTS: Search date: _____

Prior art: (e.g. Patent Document #)	Seems to disclose, which is of interest to this topic:	... but not:

POSSIBLE TO PATENT: (See Appendix B)

Ideas in this canvas that do not seem to be disclosed in the prior art:

PATENT ANALYSIS FORM

For topic of: _____ Canvas #: _____

PRELIMINARY PRIOR ART SEARCH: (See Appendix A)

Keywords: _____

Other keywords learned from early search results: _____

SEARCH RESULTS: Search date: _____

Prior art: (e.g. Patent Document #)	Seems to disclose, which is of interest to this topic:	… but not:

POSSIBLE TO PATENT: (See Appendix B)

Ideas in this canvas that do not seem to be disclosed in the prior art:

PATENT ANALYSIS FORM

For topic of: _____ Canvas #: _____

PRELIMINARY PRIOR ART SEARCH: (See Appendix A)

Keywords: _____

Other keywords learned from early search results: _____

SEARCH RESULTS: Search date: _____

Prior art: (e.g. Patent Document #)	Seems to disclose, which is of interest to this topic:	... but not:

POSSIBLE TO PATENT: (See Appendix B)

Ideas in this canvas that do not seem to be disclosed in the prior art:

PATENT ANALYSIS FORM

For topic of: _____ Canvas #: _____

PRELIMINARY PRIOR ART SEARCH: (See Appendix A)

Keywords: _____

Other keywords learned from early search results: _____

SEARCH RESULTS: Search date: _____

Prior art: (e.g. Patent Document #)	Seems to disclose, which is of interest to this topic:	... but not:

POSSIBLE TO PATENT: (See Appendix B)

Ideas in this canvas that do not seem to be disclosed in the prior art:

PATENT ANALYSIS FORM

For topic of: _____ Canvas #: _____

PRELIMINARY PRIOR ART SEARCH: (See Appendix A)

Keywords: _____

Other keywords learned from early search results: _____

SEARCH RESULTS: Search date: _____

Prior art: (e.g. Patent Document #)	Seems to disclose, which is of interest to this topic:	… but not:

POSSIBLE TO PATENT: (See Appendix B)

Ideas in this canvas that do not seem to be disclosed in the prior art:

PATENT ANALYSIS FORM

For topic of: _____ Canvas #: _____

PRELIMINARY PRIOR ART SEARCH: (See Appendix A)

Keywords: _____

Other keywords learned from early search results: _____

SEARCH RESULTS: Search date: _____

Prior art: (e.g. Patent Document #)	Seems to disclose, which is of interest to this topic:	... but not:

POSSIBLE TO PATENT: (See Appendix B)

Ideas in this canvas that do not seem to be disclosed in the prior art:

PATENT ANALYSIS FORM

For topic of: _____ Canvas #: _____

PRELIMINARY PRIOR ART SEARCH: (See Appendix A)

Keywords: _____

Other keywords learned from early search results: _____

SEARCH RESULTS: Search date: _____

Prior art: (e.g. Patent Document #)	Seems to disclose, which is of interest to this topic:	... but not:

POSSIBLE TO PATENT: (See Appendix B)

Ideas in this canvas that do not seem to be disclosed in the prior art:

PATENT ANALYSIS FORM

For topic of: _____ Canvas #: _____

PRELIMINARY PRIOR ART SEARCH: (See Appendix A)

Keywords: _____

Other keywords learned from early search results: _____

SEARCH RESULTS: Search date: _____

Prior art: (e.g. Patent Document #)	Seems to disclose, which is of interest to this topic:	... but not:

POSSIBLE TO PATENT: (See Appendix B)

Ideas in this canvas that do not seem to be disclosed in the prior art:

PATENT ANALYSIS FORM

For topic of: _____ Canvas #: _____

PRELIMINARY PRIOR ART SEARCH: (See Appendix A)

Keywords: _____

Other keywords learned from early search results: _____

SEARCH RESULTS: Search date: _____

Prior art: (e.g. Patent Document #)	Seems to disclose, which is of interest to this topic:	... but not:

POSSIBLE TO PATENT: (See Appendix B)

Ideas in this canvas that do not seem to be disclosed in the prior art:

PATENT ANALYSIS FORM

For topic of: _____ Canvas #: _____

PRELIMINARY PRIOR ART SEARCH: (See Appendix A)

Keywords: _____

Other keywords learned from early search results: _____

SEARCH RESULTS: Search date: _____

Prior art: (e.g. Patent Document #)	Seems to disclose, which is of interest to this topic:	... but not:

POSSIBLE TO PATENT: (See Appendix B)

Ideas in this canvas that do not seem to be disclosed in the prior art:

PATENT ANALYSIS FORM

For topic of: _____ Canvas #: _____

PRELIMINARY PRIOR ART SEARCH: (See Appendix A)

Keywords: _____

Other keywords learned from early search results: _____

SEARCH RESULTS: Search date: _____

Prior art: (e.g. Patent Document #)	Seems to disclose, which is of interest to this topic:	... but not:

POSSIBLE TO PATENT: (See Appendix B)

Ideas in this canvas that do not seem to be disclosed in the prior art:

PATENT ANALYSIS FORM

For topic of: _____ Canvas #: _____

PRELIMINARY PRIOR ART SEARCH: (See Appendix A)

Keywords: _____

Other keywords learned from early search results: _____

SEARCH RESULTS: Search date: _____

Prior art: (e.g. Patent Document #)	Seems to disclose, which is of interest to this topic:	... but not:

POSSIBLE TO PATENT: (See Appendix B)

Ideas in this canvas that do not seem to be disclosed in the prior art:

PATENT ANALYSIS FORM

For topic of: _____ Canvas #: _____

PRELIMINARY PRIOR ART SEARCH: (See Appendix A)

Keywords: _____

Other keywords learned from early search results: _____

SEARCH RESULTS: Search date: _____

Prior art: (e.g. Patent Document #)	Seems to disclose, which is of interest to this topic:	... but not:

POSSIBLE TO PATENT: (See Appendix B)

Ideas in this canvas that do not seem to be disclosed in the prior art:

PATENT ANALYSIS FORM

For topic of: _____ Canvas #: _____

PRELIMINARY PRIOR ART SEARCH: (See Appendix A)

Keywords: _____

Other keywords learned from early search results: _____

SEARCH RESULTS: Search date: _____

Prior art: (e.g. Patent Document #)	Seems to disclose, which is of interest to this topic:	... but not:

POSSIBLE TO PATENT: (See Appendix B)

Ideas in this canvas that do not seem to be disclosed in the prior art:

PATENT ANALYSIS FORM

For topic of: _____ Canvas #: _____

PRELIMINARY PRIOR ART SEARCH: (See Appendix A)

Keywords: _____

Other keywords learned from early search results: _____

SEARCH RESULTS: Search date: _____

Prior art: (e.g. Patent Document #)	Seems to disclose, which is of interest to this topic:	... but not:

POSSIBLE TO PATENT: (See Appendix B)

Ideas in this canvas that do not seem to be disclosed in the prior art:

PATENT ANALYSIS FORM

For topic of: _____ Canvas #: _____

PRELIMINARY PRIOR ART SEARCH: (See Appendix A)

Keywords: _____

Other keywords learned from early search results: _____

SEARCH RESULTS: Search date: _____

Prior art: (e.g. Patent Document #)	Seems to disclose, which is of interest to this topic:	... but not:

POSSIBLE TO PATENT: (See Appendix B)

Ideas in this canvas that do not seem to be disclosed in the prior art:

PATENT ANALYSIS FORM

For topic of: _____ Canvas #: _____

PRELIMINARY PRIOR ART SEARCH: (See Appendix A)

Keywords: _____

Other keywords learned from early search results: _____

SEARCH RESULTS: Search date: _____

Prior art: (e.g. Patent Document #)	Seems to disclose, which is of interest to this topic:	… but not:

POSSIBLE TO PATENT: (See Appendix B)

Ideas in this canvas that do not seem to be disclosed in the prior art:

PATENT ANALYSIS FORM

For topic of: _____ Canvas #: _____

PRELIMINARY PRIOR ART SEARCH: (See Appendix A)

Keywords: _____

Other keywords learned from early search results: _____

SEARCH RESULTS: Search date: _____

Prior art: (e.g. Patent Document #)	Seems to disclose, which is of interest to this topic:	... but not:

POSSIBLE TO PATENT: (See Appendix B)

Ideas in this canvas that do not seem to be disclosed in the prior art:

PATENT ANALYSIS FORM

For topic of: _____ Canvas #: _____

PRELIMINARY PRIOR ART SEARCH: (See Appendix A)

Keywords: _____

Other keywords learned from early search results: _____

SEARCH RESULTS: Search date: _____

Prior art: (e.g. Patent Document #)	Seems to disclose, which is of interest to this topic:	... but not:

POSSIBLE TO PATENT: (See Appendix B)

Ideas in this canvas that do not seem to be disclosed in the prior art:

PATENT ANALYSIS FORM

For topic of: _____ Canvas #: _____

PRELIMINARY PRIOR ART SEARCH: (See Appendix A)

Keywords: _____

Other keywords learned from early search results: _____

SEARCH RESULTS: Search date: _____

Prior art: (e.g. Patent Document #)	Seems to disclose, which is of interest to this topic:	... but not:

POSSIBLE TO PATENT: (See Appendix B)

Ideas in this canvas that do not seem to be disclosed in the prior art:

PATENT ANALYSIS FORM

For topic of: _____ Canvas #: _____

PRELIMINARY PRIOR ART SEARCH: (See Appendix A)

Keywords: _____

Other keywords learned from early search results: _____

SEARCH RESULTS: Search date: _____

Prior art: (e.g. Patent Document #)	Seems to disclose, which is of interest to this topic:	... but not:

POSSIBLE TO PATENT: (See Appendix B)

Ideas in this canvas that do not seem to be disclosed in the prior art:

PATENT ANALYSIS FORM

For topic of: _____ Canvas #: _____

<u>PRELIMINARY PRIOR ART SEARCH:</u> (See Appendix A)

Keywords: _____

Other keywords learned from early search results: _____

SEARCH RESULTS: Search date: _____

Prior art: (e.g. Patent Document #)	Seems to disclose, which is of interest to this topic:	… but not:

<u>POSSIBLE TO PATENT:</u> (See Appendix B)

Ideas in this canvas that do not seem to be disclosed in the prior art:

PATENT ANALYSIS FORM

For topic of: _____ Canvas #: _____

PRELIMINARY PRIOR ART SEARCH: (See Appendix A)

Keywords: _____

Other keywords learned from early search results: _____

SEARCH RESULTS: Search date: _____

Prior art: (e.g. Patent Document #)	Seems to disclose, which is of interest to this topic:	... but not:

POSSIBLE TO PATENT: (See Appendix B)

Ideas in this canvas that do not seem to be disclosed in the prior art:

108

PATENT ANALYSIS FORM

For topic of: _____ Canvas #: _____

PRELIMINARY PRIOR ART SEARCH: (See Appendix A)

Keywords: _____

Other keywords learned from early search results: _____

SEARCH RESULTS: Search date: _____

Prior art: (e.g. Patent Document #)	Seems to disclose, which is of interest to this topic:	… but not:

POSSIBLE TO PATENT: (See Appendix B)

Ideas in this canvas that do not seem to be disclosed in the prior art:

PATENT ANALYSIS FORM

For topic of: _____ Canvas #: _____

PRELIMINARY PRIOR ART SEARCH: (See Appendix A)

Keywords: _____

Other keywords learned from early search results: _____

SEARCH RESULTS: Search date: _____

Prior art: (e.g. Patent Document #)	Seems to disclose, which is of interest to this topic:	… but not:

POSSIBLE TO PATENT: (See Appendix B)

Ideas in this canvas that do not seem to be disclosed in the prior art:

PATENT ANALYSIS FORM

For topic of: _____ Canvas #: _____

PRELIMINARY PRIOR ART SEARCH: (See Appendix A)

Keywords: _____

Other keywords learned from early search results: _____

SEARCH RESULTS: Search date: _____

Prior art: (e.g. Patent Document #)	Seems to disclose, which is of interest to this topic:	… but not:

POSSIBLE TO PATENT: (See Appendix B)

Ideas in this canvas that do not seem to be disclosed in the prior art:

PATENT ANALYSIS FORM

For topic of: _____ Canvas #: _____

PRELIMINARY PRIOR ART SEARCH: (See Appendix A)

Keywords: _____

Other keywords learned from early search results: _____

SEARCH RESULTS: Search date: _____

Prior art: (e.g. Patent Document #)	Seems to disclose, which is of interest to this topic:	… but not:

POSSIBLE TO PATENT: (See Appendix B)

Ideas in this canvas that do not seem to be disclosed in the prior art:

PATENT ANALYSIS FORM

For topic of: _____ Canvas #: _____

PRELIMINARY PRIOR ART SEARCH: (See Appendix A)

Keywords: _____

Other keywords learned from early search results: _____

SEARCH RESULTS: Search date: _____

Prior art: (e.g. Patent Document #)	Seems to disclose, which is of interest to this topic:	… but not:

POSSIBLE TO PATENT: (See Appendix B)

Ideas in this canvas that do not seem to be disclosed in the prior art:

PATENT ANALYSIS FORM

For topic of: _____ Canvas #: _____

PRELIMINARY PRIOR ART SEARCH: (See Appendix A)

Keywords: _____

Other keywords learned from early search results: _____

SEARCH RESULTS: Search date: _____

Prior art: (e.g. Patent Document #)	Seems to disclose, which is of interest to this topic:	… but not:

POSSIBLE TO PATENT: (See Appendix B)

Ideas in this canvas that do not seem to be disclosed in the prior art:

PATENT ANALYSIS FORM

For topic of: _____ Canvas #: _____

PRELIMINARY PRIOR ART SEARCH: (See Appendix A)

Keywords: _____

Other keywords learned from early search results: _____

SEARCH RESULTS: Search date: _____

Prior art: (e.g. Patent Document #)	Seems to disclose, which is of interest to this topic:	... but not:

POSSIBLE TO PATENT: (See Appendix B)

Ideas in this canvas that do not seem to be disclosed in the prior art:

PATENT ANALYSIS FORM

For topic of: _____ Canvas #: _____

PRELIMINARY PRIOR ART SEARCH: (See Appendix A)

Keywords: _____

Other keywords learned from early search results: _____

SEARCH RESULTS: Search date: _____

Prior art: (e.g. Patent Document #)	Seems to disclose, which is of interest to this topic:	... but not:

POSSIBLE TO PATENT: (See Appendix B)

Ideas in this canvas that do not seem to be disclosed in the prior art:

PATENT ANALYSIS FORM

For topic of: _____ Canvas #: _____

PRELIMINARY PRIOR ART SEARCH: (See Appendix A)

Keywords: _____

Other keywords learned from early search results: _____

SEARCH RESULTS: Search date: _____

Prior art: (e.g. Patent Document #)	Seems to disclose, which is of interest to this topic:	... but not:

POSSIBLE TO PATENT: (See Appendix B)

Ideas in this canvas that do not seem to be disclosed in the prior art:

PATENT ANALYSIS FORM

For topic of: _____ Canvas #: _____

PRELIMINARY PRIOR ART SEARCH: (See Appendix A)

Keywords: _____

Other keywords learned from early search results: _____

SEARCH RESULTS: Search date: _____

Prior art: (e.g. Patent Document #)	Seems to disclose, which is of interest to this topic:	... but not:

POSSIBLE TO PATENT: (See Appendix B)

Ideas in this canvas that do not seem to be disclosed in the prior art:

118

PATENT ANALYSIS FORM

For topic of: _____ Canvas #: _____

PRELIMINARY PRIOR ART SEARCH: (See Appendix A)

Keywords: _____

Other keywords learned from early search results: _____

SEARCH RESULTS: Search date: _____

Prior art: (e.g. Patent Document #)	Seems to disclose, which is of interest to this topic:	... but not:

POSSIBLE TO PATENT: (See Appendix B)

Ideas in this canvas that do not seem to be disclosed in the prior art:

PATENT ANALYSIS FORM

For topic of: _____ Canvas #: _____

PRELIMINARY PRIOR ART SEARCH: (See Appendix A)

Keywords: _____

Other keywords learned from early search results: _____

SEARCH RESULTS: Search date: _____

Prior art: (e.g. Patent Document #)	Seems to disclose, which is of interest to this topic:	... but not:

POSSIBLE TO PATENT: (See Appendix B)

Ideas in this canvas that do not seem to be disclosed in the prior art:

PATENT ANALYSIS FORM

For topic of: _____ Canvas #: _____

PRELIMINARY PRIOR ART SEARCH: (See Appendix A)

Keywords: _____

Other keywords learned from early search results: _____

SEARCH RESULTS: Search date: _____

Prior art: (e.g. Patent Document #)	Seems to disclose, which is of interest to this topic:	... but not:

POSSIBLE TO PATENT: (See Appendix B)

Ideas in this canvas that do not seem to be disclosed in the prior art:

APPENDIX A

ABOUT PRIOR ART SEARCHES

Your patent attorney will tell you what type of thoughts, ideas and other work do not need a patent analysis. For a patent analysis, use this journal only as your patent attorney advises.

One searches the prior art for the purpose of answering a question. Different types of questions merit different types of searches, such as: patentability (novelty) searches, patent clearance searches and patent invalidation searches. Search results rarely succeed at answering the question completely and definitively, but often a preliminary *patent* search can give you some idea of what the answer is, or likely is. Your patent attorney may recommend that you use a professional searcher, if your purpose needs and justifies an answer with a high level of certainty.

For patentability (novelty) searches, often a preliminary prior art search with an acceptable level of certainty is only a patent search performed using keywords. A usual technique for such a patent search is to a) start searching patents using keywords, b) look at the first few results, c) see what *terms* the prior art uses in this field for various items of interest, and d) repeat the search using these terms as the keywords. For your analysis, you may want to first save the titles of the results in a computer document, until you finalize the keywords. Then you may want to write the seemingly most important ones in this journal.

APPENDIX B

FOR PATENTING

An idea that does not seem to be disclosed in the prior art could be patentable. Such an idea can be to do or make something new, or to do something that is known in a different way, or both, and so on. Also, look for any unexpected applications, results, or benefits. An idea may be patentable even if it is only an improvement over what is known in the prior art.

To obtain a patent: contact a patent attorney.

For general knowledge about patents, consider the book: PATENT READY®: Introductory Book For Executives, Managers, Engineers & Others   www.patent-ready.com

NOTES

Printed in Great Britain
by Amazon